SECURITY REQUIREMENTS FOR POST-TRANSITION CUBA

G. Alexander Crowther

August 2007

Comments pertaining to this report are invited and should be forwarded to: Director, Strategic Studies Institute, U.S. Army War College, 122 Forbes Ave, Carlisle, PA 17013-5244.

All Strategic Studies Institute (SSI) monographs are available on the SSI Homepage for electronic dissemination. Hard copies of this report also may be ordered from our Homepage. SSI's Homepage address is:
www.StrategicStudiesInstitute.army.mil.

The Strategic Studies Institute publishes a monthly e-mail newsletter to update the national security community on the research of our analysts, recent and forthcoming publications, and upcoming conferences sponsored by the Institute. Each newsletter also provides a strategic commentary by one of our research analysts. If you are interested in receiving this newsletter, please subscribe on our homepage at *www.StrategicStudiesInstitute.army.mil/newsletter/newsletter.cfm.*

Document first prepared by the Strategic Studies Institute

First published by the Strategic Studies Institute:
ISBN 1-58487-300-0

This Juniper Grove edition:
ISBN 978-1-60355-086-4

FOREWORD

The United States, particularly the Army, has a long history of involvement with Cuba. This history has included the Spanish-American War of 1898, military interventions in 1906 and 1912, the 1961 Bay of Pigs Invasion, the 1962 Missile Crisis, counterinsurgency, and low intensity warfare in Latin America and Africa against Cuban supported guerrilla movements.

During the Cold War, Fidel Castro's communist takeover on January 1, 1959, heightened U.S. concerns and highlighted the threat Cuba posed as a strategic ally of the Soviet Union. The collapse of the Soviet bloc in the 1990s raised hopes for an end to the communist regime in Cuba. However, after almost 5 decades of authoritarian rule, the Cuban dictator remains firmly in power. On July 31, 2006, his brother, Raul Castro, assumed provisional presidential power after an official announcement that Fidel was ill and would undergo surgery.

This monograph is designed to contribute to the process of conceptualizing a post-Castro future for the Cuban armed forces. They will need to be integrated into the family of Western Hemisphere militaries, supporting democracy, subordinate to elected civilian leaders, and respectful of human rights. This integration will require mission and structure changes. Colonel Alex Crowther draws attention to the need to engage the Cuban military and proposes a way ahead for the military, as well as for the rest of the hemisphere.

The Strategic Studies Institute is pleased to offer this report as part of its ongoing analytical program in support of Army participation in national security policy formulation and implementation.

DOUGLAS C. LOVELACE, JR.
Director
Strategic Studies Institute

BIOGRAPHICAL SKETCH OF THE AUTHOR

GLENN ALEXANDER CROWTHER is a Research Professor of National Security Studies in the Strategic Studies Institute, U.S. Army War College. He served in a variety of platoon and company level positions in the 17th Infantry Regiment in the 7th Infantry Division (Light) at Fort Ord, California. He commanded Headquarters and Service Company, United Nations Command Security Force—Joint Security Area—Pan Mun Jom and subsequently the Cheju-do Training Center in Korea. Colonel Crowther has also served in a variety of Foreign Area Officer, Operations Officer and Strategist positions in U.S. Army South and U.S. Southern Command in Panama and Miami. He has served two tours at the Pentagon working as a Strategic Plans and Policies Officer on the Department of the Army Staff and as a Politico-Military Officer at the Joint Staff J-5 (Strategic Plans and Policies Directorate). Colonel Crowther is a graduate of the Infantry Officers Basic and Advanced Courses, the Combined Arms and Services Staff School, and the Command and General Staff College at the School of the Americas. He was also an Army War College Fellow at the Fletcher School of Law and Diplomacy at Tufts University. Colonel Crowther has a B.A. in International Relations from Tufts University, an M.S. in International Relations from Troy State University, and a Ph.D. in International Development from Tulane University.

SUMMARY

This monograph serves multiple purposes, the most important of which is to contribute to the thought process of dealing with the Fuerzas Armadas Revolucionarias of Cuba (FAR). Change is inevitable in Cuba. Both Fidel Castro and his brother Raul are aging. Their passing will trigger either a succession or a transition. With that change, Cuba's security requirements will change as well. This monograph analyzes security requirements that the new Cuba will face and proposes what missions and structure the Cuban security forces might have after a transition.

The overall long-range U.S. goal is a stable, democratic Cuba which is integrated into the global market economy. The U.S. Government Commission for Assistance to a Free Cuba says that if a Cuban government asks for assistance, the United States could be made available "in preparing the Cuban military forces to adjust to an appropriate role in a democracy."

The Cuban military will have to change with the times, altering its focus from the territorial defense of Cuba and internal security to missions that are consonant with modern circum-Caribbean militaries: control of air- and sea-space against transnational criminals. The military will need a new structure for these missions, less focused on insurgency in defense of the island and more focused on a common operating picture and integration with the efforts of Cuba's neighbors. This monograph proposes a way ahead in preparing Cuban forces for the future, integrating them into the Western Hemisphere community of militaries, and ensuring their support for democracy, subordination to elected officials, and respect for human rights. It also suggests constructive engagement of the Cuban military with the international community. This change is inevitable, and can be relatively painless or long and difficult. Both the Cuban military and the international community have to decide which way they want it to be.

SECURITY REQUIREMENTS FOR POST-TRANSITION CUBA

INTRODUCTION

The transfer of power from Fidel Castro to his brother, Raul, in August 2006 prompted the *Economist* to declare "the beginning of the end of the Castro era."[1] Although no one knows when Fidel will finally pass on the reins of power, the time is approaching. Another unknown is the type of handoff. Fidel could be succeeded by another communist regime, or there could be a transition to a different type of regime. The worst case scenario would find Cuba descending into chaos if no one could replicate Fidel's ability to hold Cuba together.

This monograph posits a change to a different type of government, and assumes that a transition government is in place and has asked for assistance from the United States, and that Cuba is ready to tackle the difficult question of where the follow-on force to the *Fuerzas Armadas Revolucionarias* (FAR, the Cuban armed forces) should go. A similar type communist regime would probably not seek to change its military as it would be a major pillar of the regime. Therefore, only a post-transition regime would desire to modify the Cuban military. When the Cuban government is ready for assistance from the United States, America needs to be prepared to provide it. This monograph offers a template for that assistance.

The Cuban military is currently designed and postured for two major missions: the defense of Cuba against U.S. aggression and the provision of internal security. A post-transition government would not be challenged by the first and should not require the military to provide internal security, so the military needs to be redesigned and reoriented.

This monograph begins with a discussion of U.S. policy towards Cuba. Unlike most countries in the world, in the case of Cuba, a web of executive policies and laws produced by the legislative branch significantly limits flexibility in dealing with both the Cuban government and the FAR. Any proposal for dealing with any government organization within Cuba must start with an analysis of U.S. policy because of these limitations.

The author next examines the FAR, including both its history and its current state. As with everything in Cuba, the post-Cold War era has had a huge impact on the FAR. A current snapshot of the FAR is therefore important. From there, the author analyzes the threats that Cuba will probably face. He examines historical threats to Cuba and current threats to the Caribbean, and also considers modern transnational threats. Based on these threats, a capabilities-based approach provides a template for a post-Castro Cuban regime. The author proposes a structure for a post-transition force based on the capabilities that a Cuban force would require.

Additionally, the author examines Western Hemisphere states, with an emphasis on the sizes and types of militaries in countries that have similar territory, size, and length of coastline. This analysis provides a benchmark that may be used to develop a rough estimate of what an appropriate size might be for a Cuban military.

The Nicaraguan experience of the 1990s is the only Western Hemisphere example of a Soviet/Cuban-style military changing after the adoption of a democratic regime. As such, an analysis of the Nicaraguan case will provide some lessons learned and some perspectives to consider prior to any recommendations.

The author concludes with recommendations on "how to get there," what Cuban security forces should look like, a way ahead for influencing the change from the current structure to a 21st century Cuban security system, and precautionary measures for handling a potentially volatile situation. Mishandling security issues transitioning Cuba could have significant negative effects upon the United States, so preparation is important.

Several challenges exist. First, the Cuban military is a founding part of the state and will not give up power easily. Second, the Cuban military is in control of a major portion of the Cuban economy, and it will be very difficult to convince it to cede this position. Third, many of the elites have come from the military, and they will be strongly motivated to maintain an interventionist stance for the military. These elites must be convinced that a more subordinate role is appropriate. Fourth, the military could form a power base from which opposition elements could challenge a post-transition government.

U.S. POLICY TOWARDS CUBA

The Bureau of Western Hemisphere Affairs at the Department of State has articulated U.S. goals for the Western Hemisphere. They include strengthening an Inter-American community formed by economic partners that are democratic, stable, and prosperous; friendly neighbors that help secure the region against terrorism and illegal drugs; and nations that work together in the world to advance shared political and economic values.[7]

The State Department has also articulated goals for Cuba. According to Deputy Assistant Secretary of State for Western Hemisphere Affairs Thomas Shannon, the United States:

> would like to see a Cuba that could reintegrate itself into the inter-American community, that could return to the Organization of American States, that could become part of the Inter-American Development Bank and could play a useful role in all the other institutions of the Inter-American system. But in order for that to happen, Cuba must have a transition to democracy because ultimately a democracy is the fundamental requirement to be a member of the OAS.[3]

The latest legislation that affects U.S. Cuban policy is the Cuban Liberty and Democratic Solidarity or Libertad Act of 1995, also known as the Helms-Burton Act. The Libertad Act has two sections that deal with the Cuban military: Section 201 states that the United States will "assist a transition government in Cuba and a democratically elected government in Cuba to prepare the Cuban military forces for an appropriate role in a democracy."[4] Section 202 states that the United States will ensure that Military Adjustment Assistance provided "to a transition government in Cuba and to a democratically elected government in Cuba shall also include assistance in preparing the Cuban military forces to adjust to an appropriate role in a democracy."[5]

Specific U.S. goals in Cuba have also been articulated by the Commission to Assist a Free Cuba (CAFC). The commission's original report of May 6, 2004,[6] provided some policy guidance on the FAR. It articulated the importance of dealing with the FAR by saying that "the military's role—if any—in a democratic Cuba will likely be a key issue" and recommended the long-term goal of eliminating conscription and transitioning the FAR to an all-volunteer force. The report also pointed out that the "Cuban security forces are profoundly politicized and serve as agents of repression" and that the "United States would recommend the dissolution of the apparatus of political repression, including the General Directorate for State Security." CAFC I also recommends removing the police functions from the military.

One of the most contentious roles that the FAR plays is within the economy. CAFC I goes into great detail on this subject, and recommends that the FAR be removed from the economic sector. This policy is in keeping with long-term U.S. interests in the hemisphere, where U.S. organizations have long sought to remove militaries from this sphere.

In its second report (July 2006) called CAFC II, the Commission said that "support for professional, institutional military" would be important to,

> promote and guarantee the professionalism, dignity, and political neutrality of their [Cuba's] armed forces. A Cuban transition government will likely rely on this institution to perform many tasks during and after the transition period. The challenge for the transition government will be to harness the military's energies and direct it [sic] in ways that contribute to a successful transition period.[7]

CAFC II recommends that "Cubans can draw from those experiences by asking former communist countries to provide defense and security experts to help as the Cuban military prepares to serve as a professional force under the authority of a democratically-elected civilian government." This government would count on an apolitical, neutral military in order to thrive. According to CAFC II, the United States "should encourage the Cuban Transition Government to focus on those steps that will allow the election of a truly democratic, representative government that can take on that historic challenge." Therefore the United States should engage the FAR as soon as possible to articulate that the standard is to act in that manner, as the other militaries in the Western Hemisphere do.

This monograph posits that most of the CAFC I and II goals have been achieved, with a Cuban military that supports democracy, respects human rights, and is subordinated to civilian authorities.

THE FAR

Civil-military relations in Cuba are not consonant with the Huntingtonian paradigm, namely that the overall goal is a professional military subordinated to civilian leadership.[8] Although the FAR is subordinated to the Castro brothers, this is a sultanistic paradigm,[9] not a modern civil-military relation as recognized throughout the Western Hemisphere.

The FAR is the Cuban military, and is subordinate to the *Ministerio de las Fuerzas Armadas Revolucionarios* (Ministry of the Revolutionary Armed Forces [MINFAR]). Unlike many militaries in communist countries, which were developed by communist parties after successful takeovers, the body that became the FAR was formed before the revolution and the party was created after the revolution. In this particular case, the FAR actually formed the basis for the party.[10] As such, they are different from typical communist-inspired militaries and must be approached differently. According to the International Institute for Strategic Studies,[11] in 2005 the FAR consisted of 75,500 active forces and 1,159,000 reserve forces between military and paramilitary organizations. Of the active duty forces, the Army consisted of 38,000; the Navy consisted of 3,000; the Air Force consisted of 8,000; and the paramilitary 26,500, which was split between the State Security (20,000) and 6,500 Border Guards (*Tropas Guarda Fronteras* [TGF]). The paramilitary forces report to the *Ministerio del Interior* (Ministry of the Interior [MININT]). Of the reserve forces, the Army had 39,000, the *Ejército Juvenil de Trabajo* (Youth Labor Army [EJT]) had 70,000, the Civil Defense Force had 50,000, and the *Milicias de Tropas Territoriales* (Territorial Militia [MTT]) consisted of approximately 1,000,000 people. These reserve forces are designed to provide strategic depth and to fight the *Guerra del Todo Pueblo*, or War of the Entire People/War of the Entire Population. This doctrine was developed in the early 1990s after the collapse of the Soviet Union and the concomitant beginning of the "Special Period." This doctrine is specifically designed to defend against attack from the United States. It calls for the FAR to arm the people and deploy into the back country to keep U.S. forces tied down and make the price for a U.S. invasion too high for the United States to tolerate. Another group is the Committees for the Defense of the Revolution (*Comités de Defensa de la Revolución* [CDR]). It was formed on September 29, 1960, to support the revolution. The committees are omnipresent, with the duty to monitor the activity of everyone in their neighborhoods and report or punish any anti-revolutionary activities.

The Army is organized into three regional command headquarters and three Army command headquarters. Western Command is headquartered in Havana, Central Command is in Matanzas, and Eastern Command is in Santiago.[12] The Army consists of nine mechanized infantry brigades, up to five armored brigades, one airborne brigade, one air defense artillery regiment, and one surface-to-air missile brigade. It also has one frontier brigade on duty around Guantánamo Bay and 14 reserve brigades.

The Navy has two regional commands at Cabañas in the west and Holquin in the east. It has patrol and coastal combatants, six mine warfare and countermeasure craft, and one logistics ship. The Navy also has some towed coastal defense artillery and two amphibious assault battalions consisting of 550 naval infantry. These assets are stationed at seven facilities spread throughout Cuba at Cabañas, Mariel, Havana, Cienfuegos, Punta Movida (near Cienfuegos), Holquin, and Nicaro. The Border Guards who report to the MININT have another 23 patrol and coastal combatants.

The Air Force has up to 103 fighter aircraft and 63 transport aircraft spread out at a variety of airfields throughout the country. The Air Force has two active fighter squadrons, based at San Antonio de los Baños and Holguin respectively, equipped with this mix of fighter aircraft. Additionally, the Air Force has 40 attack helicopters and 85 support helicopters. Most Cuban military aircraft are not operational. Cuban aviation assets are deployed to Baracoa, Camagüey, Cienfuegos, Guines, Holgüin, Havana, San Antonio de los Baños, San Julián, Sancti Spiritus, Santa Clara, and Santiago de Cuba. A majority of these bases are concentrated in the western portion of the island.

The Air Force also has 13 surface-to-air missile sites. Due to the perceived threat from the United States, there is an Air Defense Force called the Air Defense of the Revolutionary Armed Forces (*Defensa Anti-Aerea de las Fuerzas Armadas Revolucionarias* or DAAFAR) which has SA-2, SA-3, and SA-6 surface-to-air missiles and 23-

mm, 37-mm, 57-mm, 85-mm, and 100-mm antiaircraft guns in great quantities.

The EJT provides additional resources to the Cuban military. Sometimes called the *Cuarto Ejercito* or "Fourth Army" in reference to its supplementing of the three armies described above,[13] this force serves several functions: it comprises a body of trained people, it inculcates the values of the revolution, and it provides an important labor force that the state can use when necessary, as in the sugar cane harvest.

The History of the FAR.

The FAR has conceptual roots into the 19th century. It traces its lineage to the *mambí* or Cuban freedom fighters that fought against Spain in wars of independence from 1868 to 1898, when the struggles culminated in the U.S. intervention as part of the Spanish-American War. As the FAR forces were a product of their own revolution from 1956-58, they feel that they are the descendants of the *mambí*.

The FAR itself was based on the *Ejercito Rebelde*, which was the Army that Fidel used to overthrow the Batista regime in Cuba in 1959. The *Ejercito Rebelde* was in turn based on the 26th of July Movement, a political movement created by Fidel Castro and named after the date of the attack on the Moncada Barracks in 1953 that marked the beginning of Fidel's armed struggle. The 26th of July forces crossed to Cuba in the yacht, *Granma*, landing near Playa de Las Colorados in Oriente Province on December 2, 1956. This area of southeastern Cuba had been where all previous revolutions and uprisings had been centered. For the next 2 years, they fought a guerrilla war against the forces of Batista. The Communist Party of Cuba did not support Fidel until early 1958,[14] which irritated him and caused him to distrust the party. By January 1, 1959, the Batista regime had collapsed, and Fidel was victorious.

When Fidel gained power, he had to determine how to rule Cuba. The only people he trusted absolutely were his comrades from the *Ejercito Rebelde*, so he turned to them to form the backbone of the new government. They comprised the senior leadership for the MININT. After he declared himself a communist, he returned to his trusted friends and used the *Ejercito Rebelde* to take over the Communist Party in 1965.[15] Thus the FAR antedates the modern Cuban Communist Party and indeed has always dominated the party, a reversal of the norm. The party, as well as the FAR, has always been subordinated to Fidel Castro. He created the FAR, and he created the party.

The FAR successfully defended the island against the Bay of Pigs operation in April 1961 and then spent the rest of the 1960s continuing to prepare the defense of the island. From the late 1960s through the 1980s, the FAR also deployed overseas in support of fellow revolutionaries throughout Africa and the Western Hemisphere. The only time that FAR conventional personnel were known to be in direct conflict with conventional U.S. forces was during Operation URGENT FURY, the 1983 U.S. intervention in Grenada. Since the end of the Cold War and the start of the Special Period, the FAR has downsized significantly and is thought to have ceased overseas revolutionary support operations. The FAR now concentrates on two main missions: deterring the United States from invading the island and internal security. It currently cooperates with the U.S. Government in both counterdrug and immigrant operations.

A Monopoly of Violence?

One way that communist governments have operated is to create multiple security organizations to prevent the monopoly of power within one organization that could then overthrow the party. Cuba was theoretically no different. On paper, the MININT is opposite the FAR within the Cuban government. MININT controls the nonmilitary security services, including the State Security Troops, the DGI or Intelligence Service, and the TGF or Frontier Guards. These forces have always oriented on internal security.

The reality of the situation is different. As described above, Fidel looked to the *Ejercito Rebelde* to provide leadership for Cuban government services, and the leadership of MININT came from the FAR. Since the FAR is controlled by Raul Castro and the MININT is controlled by the FAR, in actuality there is a monopoly of violence in Cuba, in violation of communist doctrine. Fidel Castro never avoided violating any doctrine when it suited him. In this particular case, he has ensured that all forces are directly subordinated to him personally, through his brother, Raul. This control was de facto through 1996, when the FAR took direct, overt control of MININT.[16]

This takeover has implications for dealing with both the FAR and the MININT. The FAR has always considered itself to be from the people and thus incapable of a Tiananmen Square type operation. Indeed, the domestic record of the FAR appears to be one of strict avoidance of the use of force against the Cuban people. When one examines the far more open record of the Nicaraguan armed forces that were cast in the same mold as the FAR, the absence of violence against Nicaraguans is remarkable in a region where military violence against citizens has been the historic norm.[17]

THE CUBAN DEFENSE FORCES

A transitioned Cuba cannot continue with an unchanged FAR. It must form a new military. A new Cuban military should be based on the FAR. Designing such a force requires an analysis of current threats and a conceptualization of capabilities based on those threats. From there, armed forces can be designed to provide those capabilities.

The Threats.

Cuba has had several types of threats which can prove instructive when planning for post transition security requirements. These include traditional threats, current Caribbean threats, and transnational threats that can face any state in the 21st century. In the first category, challenges to the Cuban government have included outside intervention, civil war or uprisings, smuggling, corruption, racial tensions, and natural disasters, mainly storms. Cuba will not face outside intervention but may face an uprising and racial tensions if a transition is not handled well; however, the main historical threats that Cuba will face again will likely center on smuggling, corruption, and natural disasters.

Among the current threats to Caribbean governments are transnational crime, pollution or other environmental challenges, a lack of resources, globalization, and terrorist use of the Caribbean as a platform against the United States. Transnational crime mainly centers on illicit trafficking to include people, money, arms, and drugs. An additional specific threat that a transitional Cuban government will probably face is controlling arms and munitions.

The Capabilities.

Based on the threats discussed above, the Cuban security forces need to have a minimum set of capabilities to control their maritime space, their air space, their terrestrial space, and to respond to natural disasters, mainly hurricanes. The maritime control mission should include a search and rescue (SAR) capability. Additionally, due to the presence of international narco-terrorists, Cuba requires a counterterrorism capability. In the long run, the capability to perform peacekeeping operations would provide a way for Cuba to reintegrate into the community of Western Hemisphere states. Additionally, the military should be sufficiently large to provide internal security in case the Ministry of the Interior proved to be incapable of handling an emergency.

A New Structure for the Cuban Military.

The Strategic Level. To achieve these capabilities, the FAR needs a new structure. To provide terrestrial, maritime, and air control, the Cuban military requires forces that can operate in those areas. The current Cuban Army can form the basis for a ground force, the current Air Force and the air defense force should combine to form the basis for a future Air Force, and the current Navy and border guards can be combined to form a maritime force. Due to the need to control the Cuban maritime environment, this force should be oriented like the old Cuban *guarda-costa* of the 19th century.

Every country requires a policy-level organization that provides guidance for the armed services. Currently, MINFAR provides that guidance. Although it is theoretically a ministry of defense, it is actually an organization that provides for the personal control of the FAR by the Castro brothers. After a transition, MINFAR should change to a ministry of defense within the Huntingtonian sense, with a civilian leader who is appointed by the democratically-elected president.

Throughout the Western Hemisphere, the norm is for police forces to handle internal security and for militaries to handle external security. Militaries report to a Ministry of Defense and other security forces report to a MININT. MINFAR must divest itself from control of the MININT. A new MININT must be designed with care to ensure that it is also subordinated to civilian elected leaders. The police and other domestic security forces should stay with MININT. Any remaining military intelligence organizations should stay with MINFAR, and all other intelligence assets should be transferred to the new MININT.

One model that has proved effective is a civilian Ministry of Defense with a Joint Staff headed by a senior general officer. Another model that has utility would be a combined staff under a civilian minister. Regardless of which model is used, Cuban civilian and military personnel should immediately be offered positions at the various staff colleges throughout the hemisphere that are designed to train and educate civilians and military in security-oriented positions. Extra-hemispheric countries such as Spain, the United Kingdom, Japan, the Republic of Korea, and Australia should also be involved in educating senior civilians in the Cuban Ministry of Defense. At the same time, senior military personnel should be selected to go to staff colleges in countries that have consolidated democratic systems in order to assist them in understanding the many facets of subordination to civilian authorities.

The Cuban strategic level should also be able to provide a limited command and control capability. The Cuban military should have one national-level command center, similar in theory to the North American Aerospace Defense (NORAD) Command in Cheyenne Mountain, Colorado. This post should serve as the center where all radars and other sensors and reports are aggregated. It should have a picture of everything that happens in the aerospace and maritime regions surrounding Cuba. There should also be a backup capability in case the primary center is rendered inoperable by a natural disaster.

The Cuban common operating picture should be piped to regional command posts. The national command center can then order regional commands to execute operations in support of national sovereignty missions or in response to natural disasters. Additionally, the Cuban government should provide this common operating picture to friendly states in the Caribbean to assist with regional security initiatives. The Cuban government and security forces should be prepared to cooperate with the Bahamas, the Caymans, Haiti, Jamaica, Mexico, and the United States. Each of the services should be structured to support the national missions.

The Army. The Cuban Army should consist of four brigade-level headquarters. The Army component in each regional joint command should be one of the brigade headquarters. The fourth brigade headquarters should provide command and control for the Cuban strategic reserve and could be deployable to act as a headquarters for any overseas operation. In this, it would be similar to the recently formed Canadian Expeditionary Command, which provides command and control for all deployments outside of the country. This would form the basis for the long-range goal of providing a peacekeeping capability.

Each brigade/Army component command should have several subordinate units. Each should have one armored regiment, two motorized regiments, one light infantry regiment, and an artillery regiment. This allows the Cuban Army to retain a capability similar to what currently exists but enable the Army to divest itself of the Air Defense Artillery regiments in each Army area. In addition to the main combat arms organizations, each brigade should have the combat support and combat service support assets (military police, logistics, engineers, military intelligence, etc.) to make it into a brigade combat team capable of stand-alone activities.

The Cuban Army strategic reserve capability can consist of the fourth brigade mentioned above, with one armored regiment, one motorized regiment, and the Cuban airborne regiment. This allows a downsizing of the present strategic reserve but retains a capability.

The Army should also have a reserve component. This currently consists of 14 brigades, but can be downsized to 8 regiments, with 2 assigned to each of the 4 Army brigade headquarters.

The Army can use the assets saved by demobilizing the air defense assets and spare combat arms assets to form engineer units. Cuba will have a significant infrastructure shortfall for many years after a transition. Each of the four brigades should have a minimum of one engineer regiment. Additionally, Cuba might form some special military units designed to upgrade and repair key infrastructure such as roads, railroads, and port facilities. They could also perform support missions preparing for and recovery from natural disasters.

The Navy. The maritime force[18] should combine the two current maritime forces: the Navy and the *Tropas de Guardia Frontera*. This would reduce the bureaucratic overhead and provide centralized force generation. While the FAR can trace its roots back to the *mambí* of the late 19th century, as suggested earlier, the new Cuban maritime force could trace its roots to the very competent *guarda costa* of the colonial time period. The Navy's overall mission would be maritime control, centering on border control, customs, and smuggling prevention. This mission set calls for it to be configured for patrol and counterdrug operations, SAR, and multinational regional maritime cooperation operations. In terms of facilities, the Navy has two bases near Cienfuegos; one should be closed. The Cuban Navy should also have a mix of forces deployed from the remaining six bases.

These missions call for the Navy to run long-range patrols along the extensive Cuban coastline. There should be some long-range 110+-foot boats to patrol into the Straits of Florida; the waters to the northwest of Cuba; the Yucatán Channel; the waters to the south of Cuba; and the areas between Cuba, Jamaica, and Haiti. Twenty four of these large patrol craft should be stationed, six per base, at Mariel, Cienfuegos, Holguin, and Santiago de Cuba to cover the farther reaches of Cuban maritime space. This allows for two being on patrol, two preparing for patrol, and two in maintenance.

Small, short-range 80-foot or less patrol craft such as Boston Whalers or Rigid-hull Inflatable Boats should control the area near ports as well as performing forward patrols of the archipelagos on the north-central coast of Cuba between the mainland and Cay Sal Bank and the Bahamas, and in the archipelagos off of Piña del Rio and the Isla de Juventud off the south coast. They should be home-ported at the six remaining Navy bases but forward-deployed into patrol areas that allow for them to perform their presence missions. Based upon the size of their port facilities, Havana should have eight of these patrol craft; Santiago de Cuba, Matanzas, Cienfuegos, Nuevitos, and Mariel should have six patrol craft each; and Puerto Padre and Moa should have three patrol craft apiece.

Although the FAR currently has a type of Marine Corps, any follow-on force has no need for an amphibious capability; however a proposed *guarda costa* should have some sort of boarding capability, either through a maritime naval infantry or a SEAL-type special operations capability.

The FAR currently has a robust naval radar capability, which should be maintained or even expanded to observe the waters near shore, providing another part of the Cuban common operating picture.

The Air Force. To dominate its airspace, the Cuban Air Force[19] requires interceptors and rotary-wing assets as well as a robust radar architecture to provide the airspace portion of the common operating picture.

For their strategic airspace dominance, the Cubans depend on their former Soviet fighter aircraft, including 50 MiG 21 *Fishbed* aircraft, 20 MiG 23 *Flogger* aircraft and three MiG 29 *Fulcrum* aircraft. This aircraft mix is probably too much for the Cubans to sustain. The Air Force should concentrate on one airframe to simplify maintenance and training. Given that the Cuban military will not have to defend against fighter or bomber aircraft, the MiG 21 airframe is sufficient for any airspace control missions the Air Force needs to perform. The Cuban government should sell its MiG 23 and MiG 29 aircraft and concentrate on making its fleet of MiG 21 aircraft flyable.

The current fleet of rotary wing aircraft consists of 40 Mi-8 *Hip*, 45 Mi-17, and 40 Mi-25/35 *Hind* helicopters. This fleet will serve the Cuban government well, given the large number of small islands and large amount of territorial sea. The *Hind* helicopters can also serve in the airspace control capacity when vectored in by command and control assets.

Cuban transport aircraft consist of a variety from small AN-2 *Cub* to large IL-76 *Candid* models. The Cuban military should keep all of these in the short term to ascertain future needs. With small islands spread over a large area and a substandard transportation infrastructure on the mainland, the Cuban military will likely need all of its airlift capabilities into the foreseeable future.

This Air Force should be deployed to fields throughout Cuba to respond to commands from the three regional headquarters. In addition to combat aircraft, modest fixed-wing airlift assets would be required to provide support for international operations as well as intra-Cuba military support. This capability would also support any airborne and special operations forces. The Cuban military would also require a rotary-wing capability, which should be forward deployed with an orientation on providing a SAR capability, as well as providing air mobility to Cuban ground and maritime forces and supporting disaster relief operations.

Joint Capabilities. In order to support a counter-terrorism capability, Cuba requires special operations forces (SOF). Although it would not need a huge counterrerrorism force, Cuba could use one SOF battalion. This capability should be based on the current SOF capability within the FAR.

Many different armed forces in the Western Hemisphere have developed a peacekeeping capability. Once Cuba has transitioned, it can also form such a capability, utilizing assets from each of the four Army brigades and using detachments from the *guarda costa* and the Air Force as appropriate.

The Operational Level.

As discussed above, there are currently three Army, two Navy, and two Air Force regional headquarters. These should be consolidated. Each of the Army headquarters should change to regional joint headquarters. This is a transition that the Canadian Forces undertook in 2006 to promulgate a more joint character to their armed forces. This "Regional Joint Task Force" structure has worked well for Canada. Three Cuban joint headquarters with responsibility for terrestrial, maritime, and airspace defense could find synergies, especially in the synchronization of air and maritime forces. These headquarters would also be able to execute disaster preparedness and relief operations.

Each of the regional headquarters should have three component commands similar to service component commands for the current U.S. regional combatant commands. As an example, the Western Command would contain a Western Command Joint Headquarters with Western Army Command, Western *Guarda Costa* Command, and Western Aerospace Command. Each joint command should be equipped to provide command and control for terrestrial, maritime, and aerospace operations and be able to tap into the national-level common operating picture, so that it can determine which combination of forces should be deployed to face challenges.

How Large Should Cuban Security Forces Be?

When conceptualizing an appropriate size for the Cuban military, it may be instructive to examine some attributes that Cuba shares with other countries in the Western Hemisphere. These comparisons can provide force size guidance that might be appropriate for a post-transition Cuban military. Territory size and length of coastline are two indicators that we can use.

Territory size is an important driver for the proper sizing of a military, and most countries in the Western Hemisphere have problems with territorial control. From Honduras and Nicaragua in Central America, to Colombia and Peru in South America, to Haiti in the Caribbean, many countries have never consolidated government control over their territory.

Cuba, located astride the major drug route from the source zone of South America to the United States, is in significant danger of penetration by transnational criminals in the post-transition time period. As Haiti currently shows, a lack of government control leads to near-anarchy and a very strong narco-presence. As such, Cuba must count control of terrestrial territory as very important.

Cuba has over 110,000 square kilometers of territory,[20] the same size as Pennsylvania. It is slightly smaller than Nicaragua and Honduras and larger than Guatemala and Panama. Nicaragua has a total of 14,000 military and Honduras has about 12,000. Guatemala is decreasing to about 15,000 military. Panama has a paramilitary force that totals slightly less than 12,000.

All four countries have territory that is beyond state control. A case could be made that the militaries of these countries are too small to control their territory, so a military of around 49,000 is probably about right for Cuba.

Coastline length can be an important factor in determining the size of a naval security force, especially in the Caribbean. Any island there must be able to control its maritime environment, or transnational criminals will quickly dominate the area. Within the Western Hemisphere, Chile, Argentina, and Colombia have similar-sized coastlines. Chile has a Navy of 19,000; Argentina, 17,000; and Colombia, 22,000.[21] When compared to these countries, the 3,000 personnel and handful of watercraft that Cuba can deploy demonstrates that its maritime capability is extremely low.

Western Hemisphere Templates.

What assumptions may be derived from territory and coastline data? When one analyzes territory size, Cuba has a much larger military than the four countries that are similar in size. However, all four have significant problems with territorial control. The result is that the Cuban military should not be as small as the militaries of Nicaragua, Honduras, Guatemala, or Panama.

Cuba is the largest island in the Caribbean. As such, its coastline is much longer than those of other countries in the sub-region, and comparable to the larger countries in the hemisphere. The lesson is that Cuban naval forces are woefully inadequate to dominate the maritime space. Comparison to the navies of countries with similar coastlines indicates that the Cuban Navy should be dramatically expanded.

Although a country should not base the size of its military entirely on the sizes of the militaries of other countries with similar characteristics, this comparison can provide a "sanity check" on a proposed sizing. This comparison shows that the overall size of the Cuban military is probably correct; however the Navy should be larger.

NICARAGUA – A CASE STUDY

Cuba assisted the *Frente Sandinista de Liberation Nacional* (FSLN or Sandinistas) in their struggle against the Somoza regime in Nicaragua. When Somoza fled, the opposition formed the Provisional Government, which ended up being dominated by Daniel Ortega and the FSLN. The government formed the *Ejercito Popular Sandinista* (EPS) from the FSLN ground fighters when it faced pressure from the United States. It rapidly expanded these forces to include an Air Force, a large rotary wing fleet, tanks, and the other accoutrement of a mechanized force based on Soviet and Cuban doctrine.

The EPS spent the 1980s supporting the FMLN in El Salvador, fighting against U.S.-backed counter-revolutionaries (the *contras*) and preparing for an attack by the United States. With the Esquipulas II Central American peace accords of 1987, violence diminished rapidly, and the Nicaraguan government agreed to hold elections. The FSLN lost those elections in 1990, and Doña Violeta Chamorro became the president. This was the time for the United States to engage with the new, democratically elected government of Nicaragua. Instead, the United States isolated that government.

The new Nicaraguan government faced a huge challenge in civil-military relations. President

Chamorro inherited the EPS and was forced to accept the continuance of Humberto Ortega, one of the main FSLN actors and brother of the FSLN president that Chamorro defeated, as the *Comandante* of the EPS as part of her transition plan. She chose discretion and maintained Humberto Ortega as the Minister of Defense for several years. She did not insist on huge changes in the EPS in the short term, concentrating instead on consolidating her rule over Nicaragua. This approach proved to be very effective as the EPS eventually transferred its loyalty from the FSLN to the government of Nicaragua, rebranding itself as the *Ejercito Nicaragüense* (EN).

In the aftermath of the 1990 election, the United States linked assistance with elimination of Sandinista influence in the government in general and the military in particular. Given this atmosphere, military ties took a while to develop. The EPS/EN was very suspicious of the United States in the 1990s. Hurricane Mitch changed the situation. Nicaragua was hard hit. The United States responded with military deployments into the region to assist with recovery. Due to the dire circumstances prevalent in Nicaragua, the EN determined to work with the United States. Much to its surprise, the EN found the U.S. military a worthy partner. In response to its changed perception of the U.S. military, the EN placed representation in Washington, DC. In June 2000, it assigned a brigadier general as the first Nicaraguan Defense Attaché since the 1979 revolution. His mission was to effect a rapprochement between the EN and the U.S. military. Within 2 years, *Comandante* of the EN General Javier Carrion visited the United States and laid a wreath at the Tomb of the Unknowns at Arlington National Cemetery. Relations had been normalized.

Nicaragua and Cuba.

Parallels exist between the Nicaraguan case and Cuba. Each country is small and located near the United States in the circum-Caribbean. Each was heavily influenced by the United States. Each fought against a dictator initially supported by the United States. Each had a revolutionary form of government that was actively opposed by the United States and subsequently had unfavorable relations with the United States.

Some lessons may be learned from the post-transition situation in Nicaragua. First, animosity between the U.S. Government and elements in Nicaragua slowed improvements in relations. Second, the attitude of the EN prevented a rapprochement for 10 years. Third, cooperation between the U.S. military and the EN provided a catalyst that produced a rapid improvement in relations between the militaries. From these lessons, several conclusions can be derived. First, animosity between elements in the United States and elements in Cuba can cause bad relations between Cuba and the United States after a transition. Second, actively diminishing any animosities between actors in the two countries can have a rapid payback. Third, any work with the FAR can have a positive influence on post-transition relations.

RECOMMENDATIONS

The following 14 actions could assist reformation of the Cuban armed forces.

1. Engage the Cuban military. As mentioned earlier in this monograph, Section 201 of the Libertad Act states that the United States will "assist a transition

government in Cuba and a democratically elected government in Cuba to prepare the Cuban military forces for an appropriate role in a democracy."[22] Section 202 states that the United States will ensure that Military Adjustment Assistance provided "to a transition government in Cuba and to a democratically elected government in Cuba shall also include assistance in preparing the Cuban military forces to adjust to an appropriate role in a democracy."[23] CAFC II states that a Cuban transition government will likely rely on the military to perform many tasks during and after the transition period. The challenge for the transition government will be to harness the military's energies and direct it in ways that contribute to a successful transition period.[24] The only way to make this happen is to engage the FAR and communicate the standards that apply to militaries in the Western Hemisphere: support of democracy, subordination to civilian authority, and respect for human rights.

CAFC II said that "support for [a] professional, institutional military" would be important to "promote and guarantee the professionalism, dignity, and political neutrality of their armed forces." It also recommends that "Cubans can draw from those experiences by asking former communist countries to provide defense and security experts to help as the Cuban military prepares to serve as a professional force under the authority of a democratically-elected civilian government." This government would require an apolitical, neutral military in order to thrive.

2. Transition MINFAR to a Ministry of Defense that subordinates the Cuban armed forces to civilian authority. All actors should seek a mission change for the Cuban armed forces. In accordance with CAFC I, the Cuban government should remove police

functions from the military, as well as removing the Cuban military from internal security missions. The Cuban government should form a Ministry of Defense modeled after successful Western Hemisphere models such as Chile. This effort will require support from several sources, including providing education at the Center for Hemispheric Defense Studies at Fort McNair and assistance from other countries which have made this transition, including Spain, Chile, Brazil, and Argentina. Assistance from the United Kingdom, Canada, Japan, Germany, and Australia could be helpful as well. In addition, overseas Cubans could assist. As CAFC II says, "Cubans abroad with military and defense backgrounds could be invited by the Cuban Transition Government to form teams to train, mentor, and advise Cuban defense and military leadership through the transition."

3. Separate MINFAR and MININT. Normally internal and external security missions belong to separate organizations. As described earlier, the MINFAR has no desire to be involved in internal security operations; however it supervises the MININT, which maintains the responsibility. MININT should be removed from the control of MINFAR and transferred to another governmental organization, such as a MININT. The police and other domestic security forces should stay with MININT. Although there may be a need for some sort of military intelligence, all other intelligence assets should be transferred from MINFAR control.

4. Adopt a long-range goal of removing the FAR from the economy. It will be very difficult to wean it from the money that the FAR makes from tourism and other parts of the economy. MINFAR needs to understand that its role in the economy is unacceptable. Removing MINFAR from the economy will require

a fully funded resource stream and a fully funded pension fund. A follow-on Cuban government and anyone who assists it will need to provide a budget as well as a retirement plan. These are the two greatest excuses the FAR would have to maintain its grip on the economy.

5. Do not demobilize the FAR. An attempt to demobilize the FAR could succeed, but the resultant security vacuum would produce a situation like Haiti, where the lack of security capability has left large amounts of the country to transnational criminals. If the FAR felt that an attempt to demobilize it was an attack on the country of Cuba, this would trigger the *Guerra del Todo Pueblo*, which would result in a situation where elements of society and demobilized military personnel with access to large stockpiles of weapons combine to run a full-scale insurgency aimed against the United States and the new Cuban government.

Additionally, a follow-on Cuban government will need to rely on the FAR. As CAFC I points out, "loyal and dependable military units will be needed at least until a democratic government can be consolidated and a new constitution approved by the people. Reliable military forces could help transition authorities prevent massive seaborne migration and distribute humanitarian assistance." Therefore it would be in U.S. national interests to avoid demobilization of the FAR.

6. Involve other countries and organizations in the new Cuba. The State Department cogently points out that democratic transition in Cuba is "a project of the Americas, and it's a project of a community of nations that is committed to democracy." CAFC II states that "other democratic countries or international organizations may be able to provide similar expertise

and logistical support for the military and security services." Clearly other countries need to be involved in assisting Cuba with the transition and with the professionalization of the Cuban military. Center-left governments and their militaries can assist with this project, to include Spain, Chile, Brazil, and Argentina, as well as other countries such as Colombia, Mexico, and El Salvador.

7. Insist on a major doctrinal change with the abandonment of the *Guerra del Todo del Pueblo.* This doctrine is based on the perceived need to defend Cuba from the United States. As the United States is not a threat to Cuba, this outdated doctrine should be dropped immediately. The assets dedicated to support the *Guerra del Todo del Pueblo* should be realigned to assist with the transition to a modernized military.

8. Disband the militia and the *Comites de Defensa de la Revolucion.* This is another structure change that should emanate from a transition to civilian leadership. The CDRs will have no reason to exist once a perceived threat from the United States is eliminated and should be taken apart as soon as possible. The huge militia system of the FAR should also be stood down. With the removal of the doctrine of the *Guerra del Todo del Pueblo,* there is no need for the capacity to radically expand the size of the FAR in case of emergency. As discussed previously, a federal reserve force should be developed to provide the FAR a modest expansion capability.

9. Eliminate conscription and transition the FAR to an all-volunteer force. CAFC I recommends these as long-term goals. Although this may take some time to phase in, most countries in the hemisphere have changed to this model.

10. Be patient. Allow time for change. This series of seismic changes will take a while. Although some

FAR officers will support these proposed changes, the entire group of changes will be opposed by the majority of FAR officers. Partner countries need to allow time for change. As discussed, it took a decade after elections for the Nicaraguan Army to post an attaché to Washington, DC, and another year before the United States put a Military Group into Managua.

11. Put a U.S. attaché and security assistance organization in Havana as soon as is feasible within the strictures of U.S. policy. Ask for a Cuban attaché to be posted to Washington, DC, at the same time. Although some actors will not want to move quickly, the Nicaraguan case demonstrates that, once communications are open, it can become much easier for the two sides to find common ground.

12. Provide security assistance as soon as possible. This assistance should be concentrated on international military education and training (IMET). The sooner that the Cuban military sees the U.S. backing rhetoric with tangible support, the quicker it will cooperate. Placing Cuban officers into U.S. military schools and exposing them to U.S. culture will garner tangible results as the Cubans realize that the United States is not inherently evil.

13. Do not adopt punitive, Helms-like legislation such as was passed against Nicaragua. This type of legislation will prevent the U.S. Government from cooperating with and assisting a follow-on Cuban government at the exact time when the Cuban government will need support. Punitive legislation will only drive Cuba into the arms of Venezuela and the People's Republic of China.

14. Be magnanimous. Do not seek to humiliate Cuban security forces. A little understanding can be of tremendous assistance during a time of transition.

CONCLUSION

Change is inevitable in Cuba. One aspect of that change is the FAR. The FAR currently has a doctrine designed to defend Cuba against attack from the United States. In addition, its orientation is towards internal defense. It also dominates the Cuban economy. It is subordinate directly to the Castro brothers and acts against democracy. These things will have to change. The Cuban military must be professional, politically neutral, and support democracy and democratically elected leaders. The FAR should be oriented towards control of its own sea- air- and land-space. These things are all achievable. The FAR is a force with a history of success and effort in support of the missions given by the government. As such, it is capable of change if it is called for by the Cuban leadership.

On the part of the United States, the U.S. Government needs to engage the Cuban military. At some point, the United States has had conceptual disagreements with most militaries throughout the hemisphere. With engagement and assistance, all of the militaries in the region have changed to support democracy and eschew human rights violations. With an attitude of engagement rather than confrontation, the United States could help the Cuban military to achieve the same.

ENDNOTES

1. *Economist*, August 5, 2006.

2. Department of State Western Hemisphere Affairs Web Site, *www.state.gov/p/wha/*, accessed February 12, 2007.

3. Thomas Shannon, Assistant Secretary for Western Hemisphere Affairs, Foreign Press Center Briefing Speech, "U.S.

Policy Toward Cuba," Washington, DC, August 23, 2006, *www.state.gov/p/wha/rls/rm/2006/71070.htm.*

4. Libertad Act, Title II—Assistance To A Free And Independent Cuba 22 U.S.C. 6066 Sec. 201, "Policy Toward A Transition Government And A Democratically Elected Government In Cuba."

5. Libertad Act, Title II, Sec. 202, "Authorization Of Assistance For The Cuban People."

6. CAFC original report, referred to as CAFC I, May 6, 2004, *www.cafc.gov/rpt/2004/c18165.htm.*

7. CAFC II, July 2006, *www.cafc.gov/cafc/rpt/2006/68097.htm.*

8. Samuel Huntington discusses the overall goal of civilian control of the military in Chapter Four, "Power, Professionalism, and Ideology: Civil-Military Relations in Theory," *The Soldier and the State*, Cambridge, MA: Harvard University Press, 1957, pp. 80-97. Huntington addresses civil-military relations in other types of regimes in Chapter Four, "Praetorianism and Political Decay," *Political Order in Changing Societies*, New London, CT: Yale University Press, 1968, pp 192-263.

9. Eusebio Mujal-Leon and Joshua W. Busby, "The Role of the Armed Forces in Transition—Lessons for Cuba," *Encuentro Magazine*, Spring 2003 pp. 127-132. Original article available in Spanish at *www.wws.princeton.edu/jbusby/papers/cubamil.pdf.* The authors remark that the Cuban regime is "a personalistic and charismatic dictatorship with an idiosyncratic mix of national-military, egalitarian, and residual Communist elements."

10. Hal Klepak, *Cuba's Military 1990-2005: Revolutionary Soldiers During Counter-Revolutionary Times*, New York: Palgrave MacMillan, 2005.

11. *The Military Balance 2005-2006*, London: International Institute for Security Studies (IISS), pp. 332-333.

12. Deployment information comes from Global Security at *www.globalsecurity.org/military/world/cuba/index.html.*

13. Klepak.

14. Richard Gott, *Cuba, A New History*, New Haven: Yale University Press, 2004, p. 161.

15. Klepak.

16. Department of State, *Annual Human Rights Reports*, available at *www.state.gov/g/drl/rls/hrrpt/*.

17. G. A. Crowther, "Central American Militaries, the United States and Democracy in Post-Cold War Central America," Unpublished dissertation, Tulane University, November 10, 2006.

18. The author thanks Captain Albert Lord, USN, for assistance in designing a new Cuban maritime force.

19. The author thanks Colonel George "Roberto" Doran, USAF, for assistance in designing a new Cuban Air Force.

20. Territory size is derived from the CIA Country Web Page, *www.cia.gov*.

21. Military sizes are taken from IISS, *The Military Balance*, 2005-2006 edition.

22. Libertad Act: Title II—Assistance To A Free And Independent Cuba, 22 U.S.C. 6066 Sec. 201, "Policy Toward A Transition Government And A Democratically Elected Government In Cuba."

23. Libertad Act: Title II—Assistance To A Free And Independent Cuba, 22 U.S.C. 6066 Sec. 202, "Authorization Of Assistance For The Cuban People."

24. CAFC II, July 2006, *www.cafc.gov/cafc/rpt/2006/68097.htm*.

Made in the USA
Monee, IL
04 November 2026

67377442R00024